Austin Craig Apgar

Pocket key of the birds of the northern United States

east of the Rocky Mountains

Austin Craig Apgar

Pocket key of the birds of the northern United States
east of the Rocky Mountains

ISBN/EAN: 9783741112669

Manufactured in Europe, USA, Canada, Australia, Japa

Cover: Foto ©Klaus-Uwe Gerhardt /pixelio.de

Manufactured and distributed by brebook publishing software
(www.brebook.com)

Austin Craig Apgar

Pocket key of the birds of the northern United States

NATURE STUDIES BY THE AUTHOR.

TREES OF THE NORTHERN UNITED STATES: Their Study, Description and Determination, for the Use of Schools and Private Students. 16mo, 1892, 224 pp. Over 400 Illustrations. Cloth, $1.00.

BIRDS OF THE NORTHERN UNITED STATES: Their Study, Description and Determination, for the Use of Schools and Private Students. Fully Illustrated. In Preparation.

MOLLUSKS OF THE ATLANTIC COAST OF THE UNITED STATES, SOUTH TO CAPE HATTERAS. 16mo, 1891, 100 pp. III Plates of over 60 Figures. Cloth, $1.00.

POCKET KEY OF TREES, BOTH WILD AND CULTIVATED, OF THE NORTHERN UNITED STATES, EAST OF THE ROCKY MOUNTAINS. With a Fully Illustrated Glossary of Terms. 1891, 40 pp. Cloth, 40 cents.

POCKET KEY OF THE BIRDS OF THE NORTHERN UNITED STATES, EAST OF THE ROCKY MOUNTAINS. With a Glossary of Terms. 1893. Cloth, 50 cents.

Copies of the above can be obtained, postpaid, for the prices given, by addressing

AUSTIN C. APGAR,
Trenton, N. J.

POCKET KEY

OF THE

BIRDS

OF THE

Northern United States,

EAST OF THE ROCKY MOUNTAINS.

———

BY

AUSTIN C. APGAR,

Author of " Trees of the Northern United States," " Mollusks of the Atlantic Coast," &c.

———

TRENTON, N. J.
The John L. Murphy Pub. Co., Printers.
1893.

S

INTRODUCTION.

The object of this pocket volume is to enable any one to determine the names of the birds by the plainest external parts. If birds are ever known they must be recognized by these features.

I have attempted to construct a Key that would contain a minimum of technical terms, and those used are defined in the short glossary at the end of the book.

After the supposed name is determined, it would be well to read a full description in such works as those of Dr. Jordan, Dr. Coues or Mr. Ridgway to verify the determination.

Method of using the Key. The most important caution in the use of a Key is *never* to read any statements but those to which you are directed by the letters in parenthesis. Rule. First read *all* the statements following the stars (*) at the beginning of the Key; decide which *one* of these best suits the specimen you have. At the end of the chosen one there is a letter in parenthesis ( ). Somewhere below, this letter is used two or more times. Read carefully *all* the statements following this letter; at the end of the one which most nearly states the facts about your specimen, you will again be directed by a letter to another part of the Key. Continue this process till, instead of a letter, there is a number and name. The name is that of the family to which your bird belongs. Turn to the latter part of the book where this number, in regular order, is found. Here, if there be more than one genus, another Key will lead to the genus, and, under it, still another Key, if necessary, will enable you to decide the species. The dimensions given are always in inches. For the wing the length is from the bend to the tip of the longest quill; for the bill the length is that of the top; the length of bird is from the tip of bill to end of longest tail feather, allowance being made for the length of neck.

The scientific names are according to the list of the American Ornithological Union. The vowel of the accented syllable is marked with the grave accent (ˋ) if long and the acute one (ˊ) if short.

The size, form and binding of the book is intended to render it very useful to collectors, hunters and all who take an interest in God's most brilliant, most sprightly, and most musical of

KEY TO THE FAMILIES OF BIRDS.

* Swimming birds: legs rather short; at least 3 toes either with full webbing or such membranes along the sides as to take the place of webbing. (Some long-legged birds with 3 full-webbed toes belong to the next group.) (**g.**) page 11.

* Wading birds: legs in most cases quite elongated; tibia always exserted and in most cases more or less naked below; toes frequently with more or less webbing at base, sometimes narrowly lobed along the sides. (**Z.**)

* Birds fitted neither for swimming nor wading. (**A.**)

A. With 2 toes in front and 2 behind (one species is 3-toed and has only 1 behind). (**Y.**)

A. With 2 toes permanently in front and 1 toe permanently behind; the outer toe is versatile and can be used either in front or behind; eyes directed forward instead of sidewise as in most birds; bill much hooked; owls. (**X.**)

A. With 3 toes permanently in front, and 1 toe (rarely absent) behind (one species of Falconidæ with hooked and cered bill has the outer toe versatile). (**B.**)

B. Bill hooked and with a distinct membrane (cere) at the base, extending past the nostrils. (**W.**)

B. Bill without a cere and in most cases not strongly hooked. (**O.**)

O. Hind toe short, small, inserted above the level of the others; front toes with a plain webbing at base; no soft membrane over the nostrils as in the doves and pigeons. (**V.**)

O. Hind toe inserted about on a level with the others and usually long. (**D.**)

D. Bill straight, the horny tip separated by a narrow portion from the base; nostrils opening beneath a soft swollen membrane (hard and somewhat wrinkled in mounted

D. Nostril not covered by a swollen membrane. (**E.**)

E. Bill stout, straight, longer than the head; feet with the outer and middle toes grown together for ½ their length; tarsus very short.............................XXIII. *Alcedinidæ.*

E. Bill very slender and long; the smallest of birds, less than 4 in. long; humming-birds...................XX. *Trochilidæ.*

E. Bill with the top ridge very short, but the gape wide and deep, reaching about to the eyes. (**T.**)

E. Bill not as above. (**F.**)

F. Hind claw much elongated, twice as long as that of the middle claw and straight or but little curved; inner secondaries lengthened, nearly as long as the primaries in the closed wing. (**M.**)

F. Not as above. (**G.**)

G. First primary short, never more than ⅔ as long as the longest, usually less than ½ as long; sometimes so short as to be barely noticeable on the under side of the edge of the wing. (**N.**)

G. First primary lengthened, always more than ⅔ as long as the longest quill. (**H.**)

H. Bill broad, depressed, wider than high at base, tapering to a point, which is abruptly hooked. (**K.**)

H. Bill higher than broad at base. (**I.**)

I. Bill stout at base and with the gape so angulated as to bring the corners of the mouth downward; no lobes or nicks along the cutting-edge of the bill. (**J.**)

I. Bill stout, with convex outline and with lobes or nicks near the center...XII. *Tanagridæ.*

I. Bill stout, compressed, notched and abruptly hooked near the tip; plumage olivaceous; tail without either white or yellow blotches.....................................VIII. *Vireonidæ.*

I. Bill not as above, little if at all hooked; colors in most species bright and distinctly marked.......VII. *Mniotiltidæ.*

J. Bill usually as long, or longer, than the head; ridge of bill so extended upward as to divide the frontal feathers; no notch at tip of bill or bristles at the rictus............XIV. *Icteridæ.*

J. Bill shorter than the head; ridge of bill not especially extending upward on the forehead.................XIII. *Fringillidæ.*

K. Rictal bristles absent; nostrils overhung with bristles; tail short, truncate and tipped with yellow or red; head crested ....X. *Ampelidæ.*

K. Rictal bristles numerous and long. (**L.**)

L. Tail rounded; bird less than 6 in. long, with creamy or orange-yellow....................Setophaga in VII. *Mniotiltidæ.*

L. Birds, if under 6 in. long, then not with rounded tails..........
.... XVII. *Tyrannidæ.*

M. Nostrils overhung with bristly feathers; tarsus scutellate behind; bill not very slender................XVI. *Alaudidæ.*

M. Nostrils exposed; bill very slender; tarsus not scutellate behind..VI. *Motacillidæ.*

N. Tarsus (booted) covered with a continuous plate along the front; no distinct scales except near the toes. (**S.**)

N. Tarsus (scutellate) covered with distinct rectangular scales along the front. (**O.**)

O. Bill stout, compressed, distinctly notched and hooked at tip; nostrils and rictus with bristles. (**R.**)

O. Bill not evidently hooked at tip. (**P.**)

P. Tail feathers acute pointed and somewhat stiff; bill decurved; bird under 6 in. long...............................IV. *Certhiidæ.*

P. Tail feathers rounded at tip and soft. (**Q.**)

Q. Bill long and stout; nostrils covered with bristly feathers; large birds, 10–25 in. long........................XV. *Corvidæ.*

Q. Bill slender, somewhat notched near the tip; nasal feathers directed forward and extending somewhat over the nostrils; small birds, not over 4½ in. long.................II. *Sylviidæ.*

Q. Bill rather slender; nasal feathers not directed forward over the nostrils; either small birds (4–6 in. long) with barred

quills, or large birds (8–12 in. long) with quills not
barred ....V. *Troglodytidæ*.

Q. Bill neither notched at tip nor decurved; nostrils concealed
by dense tufts of bristly feathers; birds 4–7 in. long.......
....III. *Paridæ*.

R. Large birds, 8 in. or more long; tail longer than the wings ..
....IX. *Laniidæ*.

R. Small birds, less than 7 in. long; tail shorter than the wings..
....VIII. *Vireonidæ*.

S. Small birds, less than 5 in. long; a bright yellow or red
crown patch..II. *Sylviidæ*

S. Birds over 5 in. long..I. *Turdidæ*.

T. Middle toe much longer than the side ones; its claw (pectinated) with minute saw-like teeth on its inner edge; plumage
soft ....XVIII. *Caprimulgidæ*.

T. Claw of middle toe not pectinated.   (**U.**)

U. Tail rounded, of stiff feathers with spinous shafts extending
beyond the webs..............................XIX. *Micropodidæ*.

U. Tail never rounded, often much forked and without spinous
tips...XI. *Hirundinidæ*.

V. Tarsus never provided with spines in the male; head usually
fully feathered, never entirely naked; game birds, ours all
under 20 in. long..............................XXX. *Tetraonidæ*.

V. Tarsus in the adult male armed with a spur; domestic fowl
and pheasants; our only wild species, the turkey, over 30
in. long..XXXI. *Phasianidæ*.

W. Head entirely naked or covered with down instead of
feathers; hind toe short, elevated; claws small.............
....XXVIII. *Cathartidæ*.

W. Head nearly or quite fully feathered; hind toe not elevated
and with a large and strong claw......XXVII. *Falconidæ*.

X. Middle claw (pectinated) with a saw-like ridge on the inner
edge; inner toe as long as the middle toe...XXVI. *Strigidæ*.

X. Middle claw not pectinated; inner toe not as long as the middle toe...XXV. *Bubonidæ.*

Y. Bill stout and decidedly hooked; parrots....................... .... XXIV. *Psittacidæ.*

Y. Bill stout and straight; tail feathers stiff and acute pointed.. .... XXI. *Picidæ.*

Y. Bill slender and curved downward; tail long, of round-tipped, soft feathers............................XXII. *Cuculidæ.*

Z. Head with a horny shield on the forehead, in other respects fully feathered...................................XXXIX. *Rallidæ.*

Z. Head with more or less naked tracts, either at the lores or around the eyes, in some the head is nearly all naked. (**e.**)

Z. Head fully feathered. (**a.**)

a. Bill hard, not sensitive, usually little longer than the head, sometimes shorter; if much longer, then compressed and very blunt at tip. (**d.**)

a. Bill weak, flexible, often long and slender, if short usually pigeon-like; gape extending about to the base of culmen; nostrils slit-like, surrounded by soft skin. (**b.**)

b. Toes with lobed membranes along their edges; tarsus much compressed; body depressed; small birds, 6–10 in. long...... ....XXXVII. *Phalaropodidæ.*

b. Toes without lobes, sometimes more or less webbed. (**o.**)

o. Bill usually shorter than the head, pigeon-like, the soft base separated by a narrowed portion from the hard tip; toes only 3, all in front, except one species has a minute hind toe; tarsus reticulate; birds, 6–12 in. long.................... .... XXXIV. *Charadriidæ.*

o. Legs exceedingly long; tarsus more than twice the length of the middle toe and claw; bill much longer than the head and more or less curved upward; birds over 12 in. long.... .... XXXVI. *Recurvirostridæ.*

o. Not as above; tarsus scutellate in front; bill slender, with a blunt tip; nostrils narrow, exposed slits in deep grooves

extending from ½ to nearly the full length of the bill.......
....XXXV. *Scolopacidæ.*

d. Bill much longer than the head, nearly straight, much compressed and truncate at tip; toes 3, webbed at base; tarsus reticulate.....................................XXXII. *Hæmatopodidæ.*

d. Bill either distinctly decurved or else shorter than the head; middle front toe and claw longer than the tarsus; wings short and rounded.................................XXXIX. *Rallidæ.*

d. Middle front toe and claw not longer than the tarsus; wings long and pointed.............................XXXIII. *Aphrizidæ.*

e. Middle claw (pectinated) with a saw-like ridge on the inner side; bill straight, acute and with sharp cutting-edges......
.... XL. *Ardeidæ.*

e. Middle claw not pectinated. (**f.**)

f. Bill very broad and flattened, widening towards the tip... ......
....XLIII. *Plataleidæ.*

f. Bill narrow, about as wide as high, gradually and decidedly curved downward for its whole length.........XLII. *Ibididæ.*

f. Bill narrow, straight for ½ its length and then decurved.........
.... XLI. *Ciconiidæ.*

f. Bill not as broad as high, about straight, not very acute; very large birds, over 40 in. long, with very long necks and legs .
.... XXXVIII. *Gruidæ.*

f. Bill about straight, soft, much grooved, as long as the head; bird 10–12 in. long.............Calidris in XXXV. *Scolopacidæ.*

g. With 4 toes all connected together with full webbing. (**m.**)

g. With only 3 front toes connected together with full webbing, or else bordered by membranes from the webbed base to the claws; hind toe rather short, sometimes wanting. (**h.**)

h. Bill with teeth or fringe-like lamellæ along the edges............
....XLIV. *Anatidæ.*

h. Bill with cutting-edges even. (**i.**)

i. Legs inserted so far back along the body that the bird, in

i. Legs so inserted that the body, in standing, takes nearly a horizontal position. (**j.**)

j. Nostrils tubular; bill made up of several pieces separated by deep grooves...L. *Procellariidæ.*

j. Nostrils not tubular. (**k.**)

   k. Upper mandible decidedly hooked at tip and plainly made up of separate pieces, one forming a kind of roof to the nostrils...LI. *Stercorariidæ.*

   k. Upper mandible plainly covered with one piece, and at least as long as the lower mandible; no horny shield on the forehead ....LII. *Laridæ.*

   k. Lower mandible longer than the upper one, and both decidedly compressed and knife-like...........LIII. *Rynchopidæ.*

   k. Upper mandible with a horny shield on the forehead; toes with lobes along the sides instead of webbing..................
....XXXIX. *Rallidæ.*

l. Front toes with broad membranes along the sides, more or less webbed at base; claws broad and flat, resembling human nails; no tail feathers.............................LVI. *Podicipidæ.*

l. Three front toes full-webbed; hind toe present, short; bill 2 in. or more long; large birds, 20 in. or more long; tail short..LIV. *Urinatoridæ.*

l. Three front toes full-webbed; no hind toe; our species are birds less that 20 in. long and have bills less than 2 in. long; tail short..LV. *Alcidæ.*

   m. Bill straight or slightly curved. (**o.**)

   m. Upper mandible decidedly hooked at tip. (**n.**)

n. Tail long (14–20 in.) and forked for ½ its length............ ...
....XLIX. *Fregatidæ.*

n. Tail moderate in length (5–10 in.), rounded, much longer than the short (3 in. or less) bill; plumage dark......................
.... XLVII. *Phalacrocoracidæ.*

n. Tail much shorter than the long (over 9 in.) and large-pouched

o. Bill stout at base, long (over 5 in.) and slightly curved near tip.. XLV. *Sulidæ.*

o. Bill slender, nearly straight, under 4 in. long; neck very long.. XLVI. *Anhingidæ.*

Family I. TURDIDÆ. (Thrushes.)

* Bill rather short, decidedly wider than high at base and conspicuously hooked at tip; tail about as long as the wings.....
....5. *Myadestes.*

* Bill longer, not depressed; tail shorter than the wings. (**A.**)

A. Plumage more or less blue........... ....1. *Sialia.*

A. Plumage not at all blue. (**B.**)

B. Bill not notched at tip; feathers in the nasal groove partly concealing the nostrils.......................2. *Hesperocichla.*

B. Bill notched at tip; nostrils exposed. (**O.**)

O. Breast with spots...4. *Turdus.*

O. Breast without spots, except in quite young birds.............
....3. *Merula.*

1. *Siàlia.*

* No chestnut or cinnamon on breast............................,...............
....*S. àrtica.* (Rocky Mountain Bluebird.)

* Some chestnut on breast. (**A.**)

A. Male with no chestnut on back.....................................
...............................*S. siàlis.* (Common Bluebird.)

A. Male with some chestnut on back................................
...........................*S. mexicàna.* (Western Bluebird.)

2. *Hesperocichla naèvia.* (Varied Thrush.)

3. *Mérula migratòria.* (American Robin.)

4. *Tùrdus.*

* Back, from forehead to tail, about uniform in color. (**B.**)

* Back not uniform in color. (**A.**)

A. Reddish on head, shading to olive on rump and tail.........
...................................*T. mustelìnus.* (Wood Thrush.)

A. Olive on head, shading to reddish on rump and tail..........
..............................*T. aorialáschkœ.* (Hermit Thrush.)
B. Back reddish from head to tail...................... ....
................................*T. fuscéscens.* (Wilson's Thrush.)
B. Back olive throughout. (**O.**)
  O. No whitish ring around the eye.................................
....*T. alíciœ.* (Gray-checked Thrush.)
  O. A whitish ring around the eye.................... ....
..........................*T. ustulátus.* (Olive-backed Thrush.)
5. *Myadéstes townséndii.* (Townsend's Solitaire.)

FAMILY II. **SYLVIIDÆ.** (OLD-WORLD WARBLERS.)

* Tarsus booted; tail feathers of nearly equal length and with no
  white markings; nostrils concealed by one or more small
  feathers ....,.....1. *Regulus.*
* Tarsus scutellate; tail feathers with white...........2. *Polioptila.*
1. *Régulus.*
  * Some black on the head; a single minute feather over the
    nostril...................*R. satrápa.* (Golden-crowned Kinglet.)
  * No black on the head; a tuft of small bristle-like feathers
    over the nostril......*R. caléndula.* (Ruby-crowned Kinglet.)
2. *Polióptila cœrúlea.* (Blue-gray Gnatcatcher.)

FAMILY III. **PARIDÆ.** (CHICKADEES, ETC.)

* Tail about as long as the wing; bill shorter than the head and
  rather stout..1. *Parus.*
* Tail much shorter than the wing; bill as long as the head and
  slender ....2. *Sitta.*
1. *Párus.*
  * Head conspicuously crested....*P. bícolor.* (Tufted Titmouse.)
  * Head without crest. (**A**)

A. Top of head brown; throat dusky.....................................
.........................*P. hudsónicus.* (Hudsonian Chickadee.)
A. Top of head and throat black. **(B.)**
B. Tail and wing feathers whitish edged................................
.....................*P. atricapíllus.* (Black-capped Chickadee.)
B. Tail and wing feathers not whitish edged........... ....
.............................*P. carolinénsis.* (Carolina Chickadee.)
2. *Sítta.*
   * White below, with some rusty brown on the crissum; cap
     glossy black, without stripes..
     ....*S. carolinénsis.* (White-breasted Nuthatch.)
   * Rusty brown below; crown black, with white stripes or leaden
     blue.....................*S. canadénsis.* (Red-breasted Nuthatch.)
   * Rusty brown or brownish white below; crown brown, without
     stripes...................*S. pusílla.* (Brown-headed Nuthatch.)

FAMILY IV. CERTHIIDÆ. (CREEPERS.)

Cérthia familiáris. (Brown Creeper.)

FAMILY V. TROGLODYTIDÆ. (WRENS AND MOCK-
ING-BIRDS.)

   * Birds over 8 in. long and with wings over 3 in. long; bill with
     bristles at the rictus. (Mocking-birds.) **(C.)**
   * Wings under 2½ in. long. (Wrens.) **(A.)**
   A. Back with black and white streaks extending lengthwise...
     ....3. *Cistothorus.*
   A. Back without streaks extending lengthwise. **(B.)**
   B. Back with some cross bars; no distinct whitish line over the
     eye... ....2. *Troglodytes.*
   B. Back without cross bars; a distinct whitish line over eye.....
     ....1. *Thryothorus.*
   C. Tail an inch longer than wings; outer tail feathers without
     distinct white blotches,......................6. *Harporhynchus.*

O. Tail but little longer than the wings.   (**D.**)

D. Color brownish; outer tail with white.................4. *Mimus.*

D. Color slate gray; lower tail coverts chestnut.....................
....5. *Galeoscoptes.*

1. *Thryóthorus.*

* Tail like the back in color, reddish brown...........................
                        T. ludovioiánus.   (Carolina Wren.)

* Tail feathers, except the middle pair, blackish.....................
....*T. bewíckii.*   (Bewick's Wren.)

2. *Troglódytes.*

* Tail and wings of about equal length................................
....*T. aëdon.*   (House Wren.)

* Tail decidedly shorter than the wings...............................
....*T. hiemális.*   (Winter Wren.)

3. *Cistothórus.*

* Bill about ⅓ in. long; crown streaked...............................
.........................*C. stelláris.*   (Short-billed Marsh Wren.)

* Bill ½ in. long; crown unstreaked.....................................
.........................*C. palústris.*   (Long-billed Marsh Wren.)

4. *M'imus polyglóttos.*   Mocking-bird.)

5 *Gáleoscóptes carolinénsis.*   (Catbird.)

6. *Harporhynchus rúfus.*   (Brown Thrush.)

FAMILY VI. **MOTACILLIDÆ.**   (PIPITS.)

Ánthus.

* Common.......................*A. pensilvánicus.*   (American Pipit.)

* Western U. S., rarely east to Minnesota............................
.................................*A. sprágueii.*   (Missouri Skylark.)

FAMILY VII. **MNIOTILTIDÆ.**   (AMERICAN WARBLERS.)

* Birds over 7 in. long; bill rather stout and compressed; wings
   shorter than the tail....................................10. *Icteria.*

* Birds under 6½ in. long; bill not stout.   (**A.**)

A. Bill depressed, broader than high at base, notched and slightly hooked; rictal bristles ½ the length of bill; length 5½ in. or less; wings longer than tail. **(G.)**

A. Bill slender, not depressed; rictal bristles small, or none. **(B.)**

B. Plumage entirely black and white streaked; no yellow anywhere; hind toe and claw very long..............1. *Mniotilta.*

B. Plumage not merely black and white; tail feathers, some or all of them, blotched with white. **(F.)**

B. Tail feathers with no white blotches. **(C.)**

C. Tail feathers with inner webs yellow and outer webs dusky; plumage chiefly yellow...........................7. *Dendroica.*

C. Tail feathers with the same color on both webs. **(D.)**

D. Conspicuously spotted below, thrush-like; back brown or dusky; legs long.......................................8. *Seiurus.*

D. Not spotted or streaked below. **(E.)**

E. Back olive or olive green; below with more or less of bright yellow; wings and tail of about equal length; tail rounded; head not striped.......................9. *Geothlypis.*

E. Crown with two black stripes separated by a broad one of buff; two other black stripes back of eyes....4. *Helmitherus.*

E. Bill very much compressed; top of bill straight, with the middle portion elevated into a distinct narrow ridge; crown not brightly striped..........................3. *Helinaia.*

E. Tail much shorter than wing; middle toe with claw decidedly shorter than the naked tarsus in front; rictus without evident bristles.........................5. *Helminthophila.*

F. Whole head and neck bright yellow; no white or yellow wing bars; bill notched, ½ in. or more long..2. *Protonotaria.*

F. Rictus with evident bristles; hind toe evidently longer than its claw; bluish gray above, with a golden-green patch on the back.. 6. *Compsothlypis.*

F. Rictus with evident bristles; hind toe scarcely longer than its claw; bill usually not very acute and usually with a slight notch near its tip.............................7. *Dendroica.*

F. Rictus without evident bristles; bill very acute, scarcely notched ....5. *Helminthophila.*

  G. Under parts bright yellow......................... 11. *Sylvania.*

  G. Belly without yellow in our species..............12. *Setophaga.*

1. *Mniotilta varia.* (Black and White Warbler.)

2. *Protonotaria citrea.* (Prothonotaria Warbler.)

3. *Helinaia swainsonii.* (Swainson's Warbler.)

4. *Helmitherus vermivorus.* (Worm-eating Warbler.)

5. *Helminthophila.*

  * Wing bars present, either white or yellow; about three of the outer tail feathers white on the inner web; head or breast with black. (**B.**)

  * Wings plain olive green or gray; no distinct white or yellow wing bars; very little, if any, white on the inner web of the tail feathers; no black anywhere. (**A.**)

  A. No colored crown patch; crissum pure white..................
...........................*H. peregrina.* (Tennessee Warbler.)

  A. Crown patch chestnut; below uniform yellow..................
...........................*H. ruficapilla.* (Nashville Warbler.)

  A. Crown patch orange brown; below greenish yellow...........
.......................*H. celata.* (Orange-crowned Warbler.)

  B. Throat and ear coverts either black, dusky, gray or olive.....
..................*H. chrysoptera.* (Golden-winged Warbler.)

  B. Throat either pure yellow or pure white, a narrow black streak back of the eye...
.....................*H. pinus.* (Blue-winged Yellow Warbler.)

6. *Compsothlypis americana.* (Blue Yellow-backed Warbler.)

7. *Dendroica* (den-drè-ca).

  * Tail feathers edged with yellow and without white; plumage chiefly yellow,.....................*D. aestiva.* (Yellow Warbler.)

  * Tail feathers blotched with white. (**A.**)

A. No wing bars, but a white blotch on the primaries near their base...
............*D. cæruléscens.* (Black-throated Blue Warbler.)

A. Primaries without white blotch; wing bars, if present, not white. (**E.**)

A. Primaries without white blotch; wing bars or wing patch white. (**B.**)

B. Rump and crown patch yellow; sides of breast also generally yellow...............*D. coronáta.* (Yellow-rumped Warbler.)

B. Rump and belly yellow; white blotches at the middle of nearly all the tail feathers; crown not yellow, usually clear ash.............................*D. maculósa.* (Magnolia Warbler.)

B. Rump and sides of neck usually yellow; bill very acute and distinctly decurved........*D. tigrìna.* (Cape May Warbler.)

B. Rump not yellow; bill not very acute. (**O.**)

O. Throat yellow or orange; crown with a small or large yellow or orange spot; outer tail feathers with outer edge white edged, as well as white blotches on the inner web................*D. bláckburniæ.* (Blackburnian Warbler.)

O. Crissum white; no bright yellow anywhere; crown black; everywhere much streaked with black...........................
.................................*D. striáta.* (Black-poll Warbler.)

O. Crissum buffy; no bright yellow anywhere; crown and throat usually chestnut..*D. castánia.* (Autumn Warbler.)

O. No bright yellow anywhere; entire upper part sky blue (male) or dull greenish, brightest on the head (female)...
....*D. cærúlea.* (Cærulean Warbler.)

O. Throat, breast and sides black or with black traces, sometimes veiled with yellowish on the tips of the feathers; outer tail feathers with outer edges, as well as blotches on the inner webs, white..
................*D. vírens.* (Black-throated Green Warbler.)

O. Not as above; throat more or less yellow. (**D.**)

D. Tail blotches oblique at the end of two or three outer feathers..................*D. vigórsii.* (Pine-creeping Warbler.)

D. Throat definitely yellow; belly white; back with no greenish.........*D. domínica.* (Yellow-throated Warbler.)

B. Wing bars and belly yellow..*D. díscolor.* (Prairie Warbler.)

B. Wing bars yellow and belly pure white............................ ....*D. pennsylvánica.* (Chestnut-sided Warbler.)

B. Wing bars brownish; white blotches square, and on only two of the outer tail feathers.................................... ....*D. palmárum* (Palm Warbler.)

B. Wing bars not conspicuous; whole under parts yellow; back ashy, without any tint of green or olive.. ....*D. kirtlandi.* (Kirtland's Warbler.)

8. *Seiùrus.*

* Crown orange-brown, with two black stripes........................ ....*S. aurocapíllus.* (Oven-bird.)

* Crown plain brownish. (**A.**)

A. Line over eye buffy; bill about ½ in. long...................... ....*S. noveboracénsis.* (Water Thrush.)

A. Line over eye white; bill about ¾ in. long...................... ....*S. motacilla.* (Louisiana Water Thrush.)

9. *Geóthlypis.*

* Wings over 2½ in. long, and decidedly longer than tail. (**B.**)

* Wings under 2½ in. long, and about equal the tail. (**A.**)

A. Wings and tail about 2 in. long.................................... ....*G. tríchas.* (Maryland Yellow-throat.)

A. Wings and tail decidedly over 2 in. long....................... ....*G. philadélphia.* (Mourning Warbler.)

B. Head without black; crown and throat ash; a whitish eye ring.............................*G. ágilis.* (Connecticut Warbler.)

B. Head with black; line over eye and under parts yellow....... ....*G. formósa.* (Kentucky Warbler.)

10. *Ictéria vírens.* (Yellow-breasted Chat.)

11. *Sylvània.*

* Tail feathers with white blotches...
.... *S. mitráta.* (Hooded Warbler.)
* No white blotches on the inner webs of tail feathers. (**A.**)
  A. Above yellow olive; crown black without streaks.............
  ....*S. pusílla.* (Green Black-capped Warbler.)
  A. Above bluish ash; crown streaked..................................
  ....*S. canadénsis.* (Canadian Warbler.)
12. *Setóphaga ruticílla.* (American Redstart.)

FAMILY VIII. **VIREONIDÆ.** (VIREOS.)
Vireo.
  * A distinct short first primary. (**C.**)
  * No distinct short first primary. (**A.**)
  A. Stout species; bill stout, blue-black; two white wing bars; a
     pale ring around eye...
    ....*V. flávifrons.* (Yellow-throated Vireo.)
  A. Slender species; bill slender, light colored; no wing bars
     nor conspicuous ring around eye. (**B.**)
  B. Crown ashy, with blackish edgings....................................
  ....*V. olivàceous.* (Red-eyed Vireo.)
  B. Crown ashy, with no blackish lines...............................
  ....*V. philadélphicus.* (Philadelphia Vireo.)
  C. Slender species; bill slender, light colored; no wing bars...
  .... *V. gilvus.* (Warbling Vireo.)
  C. Stout species; bill stout, blue black; two pale wing bars.
     (**D.**)
  D. Bird over 5 in. long; below rather whitish, with but little
     show of yellow; wings pointed and decidedly longer than
     tail..........................*V. solitàrius.* (Blue-headed Vireo.)
  D. Five in. or less long; some bright yellow below; wings
     rather rounded and but little longer than tail. (**E.**)
  E. Line around eye yellow..:......................
  ....*V. noveboracénsis.* (White-eyed Vireo.)
  E. Line around eye whitish............... *V. bélli.* (Bell's Vireo.)

FAMILY IX. LANIIDÆ. (SHRIKES.)

Lánius.

* Breast with distinct wavy cross lines; forehead white...........
..........................*L. boreális.* (Great Northern Shrike.)
* Breast with no distinct wavy cross lines; forehead black.......
.........................*L. ludoviciánus.* (Loggerhead Shrike.)

FAMILY X. AMPELIDÆ. (CHATTERERS.)

Ámpelis.

* Crissum chestnut; wing bar white................................
...............................*A. gárrulus.* (Bohemian Waxwing.)
* Crissum white; no wing bar.....................................
..................................*A. cedrórum.* (Cedar Waxwing.)

FAMILY XI. HIRUNDINIDÆ. (SWALLOWS.)

* Tail even; first primary not rough on the outer edge; plumage
    lustrous ....2. *Petrochelidon.*
* Tail nearly even; first primary with rough outer edge;
    plumage not lustrous...........................6. *Stelgidopteryx.*
* Tail decidedly forked. (**A.**)
  A. Tail forked for more than half its length; tip of outer
      feathers very narrow; color steel blue.........3. *Chelidon.*
  A. Tail forked for less than half its length. (**B.**)
  B. Plumage plain brown, without lustre................5. *Clivicola.*
  B. Plumage lustrous blue black throughout (male), whitish and
      streaky below (female)........................ ....1. *Progne.*
  B. Plumage lustrous blue green above, pure white below..........
      ....4. *Tachycineta.*

1. *Prógne súbis.* (Purple Martin.)
2. *Petrochélidon lúnifrons.* (Cliff Swallow.)
3. *Chélidon erythrogáster.* (Barn Swallow.)
4. *Tachycinéta bícolor.* (White-bellied Swallow.)

5. *Clivicola ripária* (Sand Martin.)
6. *Stelgidópteryx serripénnis.* (Rough-winged Swallow.)

FAMILY XII. **TANAGRIDÆ.** (TANAGERS.)
Piránga.
* Male red, with black wings and tail; female clear olive and yellow........................*P. erythromélas.* (Scarlet Tanager.)
* Male red throughout; female brownish olive and dull yellow.............................*P. rúbra.* (Summer Tanager.)

FAMILY XIII. **FRINGILLIDÆ.** (FINCHES.)
* Mandibles long and much curved, their points crossed at tip.. ....*4. Loxia.*
* Bill very large and stout, as high at base as long; culmen and gonys usually much curved. (**T.**)
* Bill neither very stout nor with the points crossed at tip. (**A.**)
A. Rather evenly-colored birds; there may be large patches of different colors but they are not sharply spotted or streaked either above or below; some are mottled, but not definitely so. (**P.**)
A. Decidedly spotted or streaked either above or below. (**B.**)
B. Outer tail feathers rounded at tip; middle ones very acute pointed; tail notched; hind claw elongated, nearly as long as the bill and but little curved. (**O.**)
B. About all the tail feathers narrow and acute and in many cases stiff; back conspicuously streaked; hind claw not especially elongated. (**M.**)
B. Tail feathers not especially narrow, usually rounded at tip, if pointed the ends form an obtuse angle; tail feathers not stiff. (**C.**)
C. Hind claw elongated, twice as long as the middle claw and much curved; plumage white or white and brown......... ....*10. Plectrophenax.*

C. Hind claw not twice as long as the middle claw. (**D.**)

D. Sparrows of rather large size, 7 in. or more long. (**L.**)

D. Sparrows less than 7 in. long. (**E.**)

E. Tertiary wing quills very much elongated; a white wing patch ....28. *Calamospiza.*

E. Tertials not elongated; tail rounded, its outer feathers with white; hind claw short.........................15. *Chondestes.*

E. Tertials not elongated; outer tail feathers without white. (**F.**)

F. Plumage with some definite red. (**K.**)

F. Plumage with definite yellow somewhere. (**J.**)

F. With neither clear red nor yellow anywhere. (**G.**)

 G. Plumage much streaked below. (**I.**)

 G. Not streaked below in adult birds. (**H.**)

H. Tail forked; wings and tail about equal in length; crown chestnut in adult (streaky in young)..............17. *Spizella.*

H. Tail notched; wings longer than tail; crown either ashy-brown or liver-brown; introduced birds...........9. *Passer.*

H. Tail rounded; wings either shorter than tail or of equal length; crown chestnut in adult................20. *Melospiza.*

 I. Olivaceous, no black or chestnut; bill rather stout; wings long, pointed, much longer than the notched tail; female of.. 3. *Carpodacus.*

 I. Tail rounded; back much streaked; crown with an obscure pale medium stripe.................................20. *Melospiza.*

J. With much yellow on edges of quills and tail feathers; bill quite acute...7. *Spinus.*

J. Edge of wing, line over eye, breast and part of belly yellow ....27. *Spiza.*

J. Edge of wing yellow or yellowish; breast buffy; back largely chestnut...19. *Peucæa.*

 K. Crown red; chin blackish; breast in males reddish or pinkish.................................. ....6. *Acanthis.*

K. Crown, chin, throat, and, generally, the whole plumage, with red washings; male of....................3. *Carpodacus*.

L. Head in adult striped, in young chestnut; plumage not streaked below; tail not forked...............16. *Zonotrichia*.

L. Rump, tail and wings with much chestnut or rusty red; large arrow-shaped spots on the white breast.......21. *Passerella*.

L. Bill pale; body pure white in summer; in winter, the white of body much clouded with clear, warm brown...............
....10. *Plectrophenax*.

M. Outer tail feathers with white; bend of wing chestnut......
....13. *Poocætes*.

M. Outer tail feathers without white; edge of wing yellow (**N.**)

N. Breast with yellow; throat with more or less black.............
....27. *Spiza*.

N. Breast without yellow; throat without black; plumage streaked below....................................14. *Ammodramus*.

O. Outer tail feathers almost entirely white; others, except the middle pair, tipped with black; bend of wing chestnut...
....12. *Rhynchophanes*.

O. Not as above; a brownish or chestnut-colored collar around neck...11. *Calcarius*.

P. Tail decidedly longer than the wings; large birds, over 7 in. long; black or clear brown above; sides chestnut.............
....22. *Pipilo*.

P. Tail and wings of about equal length; blackish or ashy colored above; belly and outer tail feathers white.............
....18. *Junco*.

P. Wings longer than the tail. (**Q.**)

Q. Birds over 6 in. long; wings over 3¼ in. long. (**S.**)

Q. Birds under 6 in. long; wings under 3¼ in. long. (**R.**)

R. *Male* entirely blue, or blue, red, purple, gold and white; *female* brown, with or without white.............26. *Passerina*.

R. Both sexes with crimson, black, yellow, white and plain brown..................... .... 8. *Carduelis.*

S. General color white, with some black or clear brown.........
.... 10. *Plectrophenax.*

S. Brownish above and below, with rosy edgings; black or clear ash on head; tail somewhat notched.....................
.... 5. *Leucosticte.*

S. *Male* blue, with chestnut on wings; *female* plain brown; tail even...25. *Guiraca.*

T. Conspicuously crested; plumage chiefly red or extensively washed with red; bill red; tail longer than the wings......
.... 23. *Cardinalis.*

T. No crest; tail shorter than the wings. (**U.**)

U. Rather small birds; wings 3¾ in. or less long. (**W.**)

U. Rather large birds; wings over 3¾ in. long. (**V.**)

V. Under tail coverts yellow; inner secondaries and wing coverts white; bill greenish yellow; wings nearly twice the length of the tail............................1. *Coccothraustes.*

V. General colors black and white (male) or brown streaked (female); under wing coverts rosy or yellow; tail with white blotches..24. *Habia.*

V. General colors rosy red (male) or ashy gray with brownish yellow on head and rump (female)...............2. *Pinicola.*

W. Tertiary wing feathers very much elongated; a large white wing patch......................................28. *Calamospiza.*

W. Blue (male) or brownish or tawny (female); wing bars chestnut or whitish; wings over 3 in. long..25. *Guiraca.*

W. Streaky, no yellow; *male* with much red; *female* olive brown; wings over 3 in. long ....3. *Carpodacus.*

W. Streaky, no yellow; wings white barred and under 3 in. long ....9. *Passer.*

1. *Coccothraustes* (thrés-tes) *vespertinus.* (Evening Grosbeak.)

2. *Pinicola enucleator.* (Pine Grosbeak.)

3. *Carpodacus purpureus.* (Purple Finch.)

4. *Lóxia.*
* Wings with two white bars...
.....................*L. leucóptera.* (White-winged Crossbill.)
* Wings without white.............*L. curviróstra.* (Red Crossbill.)

5. *Leucostícte tephrocótis.* (Gray-crowned Leucosticte.)

6. *Acánthis.*
* Rump distinctly streaked; inner webs of tail feathers very slightly, if at all, edged with white................................
....*A. linária.* (Redpoll Linnet.)
* Rump plain white or pinkish; tail feathers broadly edged with white............*A. hornemánnii.* (Greenland Redpoll.)

7. *Spínus.*
* Plumage not sharply streaked; no distinct ruff at base of bill..............................*S. trístis.* (American Goldfinch.)
* Plumage sharply streaked; no black on head; bill sharp and with distinct ruff at base..............*S. pínus.* (Pine Siskin.)

8. *Carduélis carduélis.* (European Goldfinch.)

9. *Pásser doméstícus.* (European House Sparrow.)

10. *Plectróphenax niválís.* (Snow Bunting.)

11. *Calcárius lappónicus.* (Lapland Longspur.)

12. *Rhynchóphanes mccównii.* (Maccown's Bunting.)

13. *Poocaétes gramíneus.* (Vesper Sparrow.)

14. *Ammódramus.*
* Outer pair of tail feathers longer than the middle pair; wings much longer than the tail. (**D.**)
* Tail distinctly double rounded; the outer pair of feathers a little shorter than the middle ones; bird about 5 in. long....
......................*A. savannárum.* (Grasshopper Sparrow.)
* Tail graduated, the outer feathers gradually shorter, the outer pair shortest. (**A.**)
  A. Tail feathers acute but not stiff; crown with a medium light stripe; inland species. (**O.**)
  A. Tail feathers acute and rather stiff; crown without medium light stripe; sea-shore species. (**B.**)

B. No yellow spot about eye..
.............................*A. caudácutus.*   (Sharp-tailed Finch.)
B. A yellow spot before eye...*A. marítimus.* (Sea-side Finch.)
C. Breast with some sharp black streaks.............................
.............................*A. henslówii.*   (Henslow's Sparrow.)
C. Breast without spots.....*A. lecónteii.*   (Le Conte's Sparrow.)
D. Back not very sharply streaked with sandy brown...............
.................................*A. prínceps.*   (Ipswich Sparrow.)
D. Back sharply streaked with blackish ....
.............................*A. sandwichénsis.*   (Savanna Sparrow.)
15. *Chondéstes grámmacus.*   (Lark Sparrow.)
16. *Zonotríchia.*
  * Head with some distinct yellow. (**B.**)
  * Plumage with no yellow anywhere. (**A.**)
  A. Crown without pale medium stripe................................
..............................*Z. quérula.*   (Harris's Sparrow.)
  A. Crown with white medium band................................
...................*Z. leucóphrys.*   (White-crowned Sparrow.)
  B. Crown stripe yellow in front...
.......................*Z. coronáta.*   (Golden-crowned Sparrow.)
  B. Crown stripe without yellow; spot over eye and edge of
      wing yellow........*Z. albicóllis.*   (White-throated Sparrow.)
17. *Spizélla.*
  * Crown grayish with a light stripe...................................
.................................*S. pállida.*   (Clay-colored Sparrow.)
  * Crown dull chestnut; no black on forehead.......................
....*S. pusílla.*.   (Field Sparrow.)
  * Crown bright chestnut. (**A.**)
  A. Forehead without black.. .....*S. montícola.*   (Tree Sparrow.)
  A. Forehead black......................*S. sociális.*   (Chippy-bird.)
18. *Júnco hyemális.*   (Snowbird.)
19. *Peucœa æstivális.*   (Pine-woods Sparrow.)
20. *Melospíza,*

* Crown with a faint pale medium stripe; bird much streaked above and on the breast and sides...................................
....*M. faciàta.* (Song Sparrow.)
* Everywhere sharply streaked; crown not chestnut...............
..................................*M. lincolni.* (Lincoln's Finch.)
* Crown bright dark chestnut; wings and tail edged with chestnut..............................*M. georgiàna.* (Swamp Sparrow.)

21. *Passerélla ilìaca.* (Fox Sparrow.)
22. *Pípilo erythrophthálmus.* (Chewink.)
23. *Cardinàlis cardinàlis.* (Cardinal Grosbeak.)
24. *Hàbia ludoviciàna.* (Rose-breasted Grosbeak.)
25. *Guiràca cærùlea.* (Blue Grosbeak.)
26. *Passerìna.*

* Gonys with a black stripe..............*P. cyànea.* (Indigo-bird.)
* Gonys without black stripe..................*P. círis.* (Nonpariel.)

27. *Spìza americàna.* (Black-throated Bunting.)
28. *Calámosplaa melanocòrys.* (Lark Bunting.)

FAMILY XIV. **ICTERIDÆ.** (BLACKBIRDS, ETC.)

* Tail feathers acute at tip. (**F.**)
* Tail feathers usually rounded, never very acute. (**A.**)
   A. Bill nearly straight, the commissure sometimes curved but the *tip* not evidently decurved. (**O.**)
   A. Bill with the *tip* evidently decurved. (**B.**)
   B. Tail much shorter than the wings, the feathers of nearly equal length..7. *Scolecophagus.*
   B. Tail as long as the wings, its feathers graduated..8. *Quiscalus.*
   O. Bill much shorter than the head, finch-like.... 2. *Molothrus.*
   O. Bill about as long as the head. (**D.**)
   D. Bill slender; wing less than 4 in. long.............. 6. *Icterus.*
   D. Bill stouter; wing 4 in. or more long. (**E.**)
   E. Side claws about as long as the middle one......................
.... 3. *Xanthocephalus.*

B. Side claws decidedly shorter than the middle one.............
.... 4. *Agelaius.*
F. Bill shorter than the head, finch-like.............1. *Dolichonyx.*
F. Bill as long as the head.............................5. *Sturnella.*
1. *Dolichonyx oryzivorus.* (Bobolink, Reedbird.)
2. *Mólothrus áter.* (Cowbird.)
3. *Xanthocéphalus xanthocéphalus.* (Yellow-headed Blackbird.)
4. *Agelaius* (aj-e-lè-us) *phœníceus.* (Red-winged Blackbird.)
5. *Sturnélla mágna.* (Meadowlark.)
6. *Icterus.*
 * Tail graduated for ¼ its length...*I. spúrius.* (Orchard Oriole.)
 * Tail nearly even...................*I. gálbula.* (Baltimore Oriole.)
7. *Scolecóphagus carolínus.* (Rusty Grackle.)
8. *Quíscalus quíscula.* (Crow Blackbird.)

FAMILY XV. **CORVIDÆ.** (CROWS AND JAYS.)

 * Tail only ⅔ as long as the pointed wings; plumage black.......
.... 1. *Corvus.*
 * Tail as long or longer than the rounded wings. (**A.**)
 A. Colors black and white; tail much graduated and much
    longer than the wings; wing over 7 in. long; no crest....
.... 2. *Pica.*
 A. Head crested; general color blue.. ....3. *Cyanocitta.*
 A. No crest; ashy gray, no blue; wings and tail about equal
    in length...4. *Perisoreus.*
1. *Córvus.*
 * Feathers of throat narrow, stiff, sharp-pointed, with distinct
    outlines ....*C. córax.* (Raven.)
 * Throat feathers short, oval, blended. (**A.**)
 A. Wing over 12 in. long..................*C. americána.* (Crow.)
 A. Wing under 12 in. long...........*C. ossifragus.* (Fish Crow.)
2. *Pica pica.* (Magpie.)
3. *Cyanocitta cristáta.* (Blue Jay.)
4. *Perisóreus canadénsis.* (Canada Jay.)

FAMILY XVI. **ALAUDIDÆ.** (LARKS.)

* First primary short; tail deeply notched................1. *Alauda.*
* First primary long; tail even or rounded.............2. *Otocoris.*
1. *Aláuda arvénsis.* (Skylark.)
2. *Otócoris alpéstris.* (Horned Lark.)

FAMILY XVII. **TYRANNIDÆ.** (FLYCATCHERS, ETC.)

* Crown with patch of red or yellow, which can be seen by dis-
placing the crown feathers. (**O.**)
. * Crown without concealed bright-colored feathers; outer
primaries not narrowed near their tips. (**A.**)
  A. Length 8 in. or more; head with a slight crest; light-
chestnut edges to the wing and tail feathers..................
.... 3. *Myiarchus.*
  A. Length less than 8 in.; no chestnut on wings and tail. (**B.**)
B. Bill all black, but little wider than high at base; wing and
tail of nearly equal length; wing 3¼ in. or more............
.... 4. *Sayornis.*
B. Bill either pale below or else the wing is less than 2¾ in.
long; bill broad and much depressed; wing 3 in. or less...
.... 6. *Empidonax.*
B. Not as above; wings over 3 in. long.................5. *Contopus.*
  O. Tail much longer than the wings and deeply forked..........
.... 1. *Milvulus.*
  O. Tail not longer than the wings and even or rounded..........
....2. *Tyrannus.*
1. *Milvulus.*
  * Cap black..............*M. tyrannus.* (Forked-tailed Flycatcher.)
  * Cap ashy...............*M. forficátus.* (Scissor-tailed Flycatcher.)
2. *Tyrannus.*
  * No olive or decided yellow; all tail feathers tipped with

* With olive and yellow; outer tail feathers with white...........
.....................................*T. verticālis.* (Arkansas Kingbird.)
3. *Myiárchus crinītus.* (Great-crested Flycatcher.)
4. *Sayórnis phoébe.* (Phœbe.)
5. *Contòpus.*
  * A tuft of white fluffy feathers on each side of rump; wing
    about 4 in. long.........*C. boreālis.* (Olive-sided Flycatcher.)
  * No conspicuous cottony tuft; wing 3½ in. or less..................
    ....*C. virens.* (Wood Pewee.)
6. *Empidònax.*
  * Distinctly yellow below...
    ....*E. flavivéntris.* (Yellow-bellied Flycatcher.)
  * Lower parts not distinctly yellow. (**A.**)
  A. A yellowish ring around eye; longest primary nearly an
    inch longer than the secondaries...............................
    ....*E. acādicus.* (Green-crested Flycatcher.)
  A. No distinct yellowish ring around eye. (**B.**)
  B. Longest primary over ½ in. longer than the secondaries......
    ....*E. pusíllus.* (Little Flycatcher.)
  B. Longest primary only ½ in. longer than the secondaries;
    tail slightly notched.......*E. mínimus.* (Least Flycatcher.)

FAMILY XVIII. **CAPRIMULGIDÆ.** (GOATSUCKERS.)

* Tail rounded; rictal bristles long and stiff.......1. *Antrostomus.*
* Tail notched; rictal bristles very short....... ......2. *Chordeiles.*
1. *Antróstomus.*
  * Rictal bristles unbranched......*A. vocíferus.* (Whip-poor-will.)
  * Rictal bristles branched..*A. carolinénsis.* (Chuckwill's Widow.)
2. *Chordeiles* (di-les) *virginiānus.* (Night Hawk.)

FAMILY XIX. **MICROPODIDÆ.** (SWIFTS.)

Chætūra pelágica. (Chimney Swift.)

FAMILY XX. **TROCHILIDÆ** (HUMMING-BIRDS.)

Tróchilus cólubris. (Ruby-throated Humming-bird.)

FAMILY XXI. **PICIDÆ.** (WOODPECKERS.)

* Head with a conspicuous crest; large birds, over 16 in. long.
  (**D.**)
* Head not crested; birds under 15 in. long. (**A.**)
  A. Hind toe single; bill broad, compressed..........4. *Picoides.*
  A. Hind toes two. (**B.**)
  B. Large birds, 12–14 in. long; belly with round black spots....
    ....7. *Colaptes.*
  B. Birds 10 in. or less long. (**C.**)
    C. Nasal groove running nearly to the tip of the bill; feathers
      with round white or black spots; no yellow..3. *Dryobates.*
    C. Nasal groove running into the tomia near the middle of the
      bill; belly with some yellow...................5. *Sphyrapicus.*
    C. Upper mandible with a ridge below the nostril extending
      to the tip of bill; the tip somewhat truncate...............
      .... 6. *Melanerpes.*
  D. Bill and nasal feathers dark colored................1. *Ceophlœus.*
  D. Bill and nasal feathers light colored............2. *Campephilus.*
1. *Ceophlóëus pileátus.* (Pileated Woodpecker.)
2. *Campéphilus principális.* (Ivory-billed Woodpecker.)
3. *Dryobátes.*
  * Back black, barred with white, not lengthwise streaked.........
    ....*D. boreális.* (Red-cockaded Woodpecker.)
  * Back black, with a long white stripe; no cross bars. (**A.**)
    A. Outer tail feathers white.. ......
      ....*D. villósus.* (Hairy Woodpecker.)
    A. Outer tail feathers black and white barred.....................
      .... *D. pubéscens.* (Downy Woodpecker.)

4. *Picoìdes.*

  * Top of head and back without white ....
  ....*P. árcticus.*   (Arctic Three-toed Woodpecker.)

  * Back with some white..
  ....*P. americánus.*   (American Three-toed Woodpecker.)

5. *Sphyrápicus várius.*   (Yellow-bellied Woodpecker.)

6. *Melanérpes.*

  * Whole head and neck crimson...
  ....*M. erythrocéphalus.*   (Red-headed Woodpecker.)

  * Whole head and neck not crimson; male with top of head
  and hind neck scarlet...
  ....*M. caroliñus.*   (Red-bellied Woodpecker.)

7. *Coláptes aurátus.*   (Flicker.)

FAMILY XXII. OUCULIDÆ. (CUCKOOS.)

Coccygus.

  * Bill yellow below; wings with reddish........... ....
  ....*C. americána.*   (Yellow-billed Cuckoo.)

  * Bill chiefly black; wings without reddish...........................
  .... *C. erythrophthálmus.*   (Black-billed Cuckoo.)

FAMILY XXIII. ALCEDINIDÆ. (KINGFISHERS.)

Céryle álcyon.   (Belted Kingfisher.)

FAMILY XXIV. PSITTACIDÆ. (PARROTS.)

Conùrus carolinénsis.   (Carolina Paroquet.)

FAMILY XXV. BUBONIDÆ. (OWLS.)

  * Tarsus nearly naked, very long, twice as long as the middle
  toe; wing under 8 in. long............................9. *Speotyto.*

A. Head with conspicuous ear tufts; plumage not chiefly white, **(D.)**

A. Head without evident ear tufts. **(B.)**

B. Plumage chiefly white; tail rounded; large bird...7. *Nyctea.*

B. Plumage not white. **(C.)**

  C. Wing 16–18 in.; tail ⅔ the length of wing......3. *Scotiaptex.*

  C. Wing 12–15 in.; tail ⅔ wing............................2. *Syrnium.*

  C. Wing 8–10 in.; tail ¾ wing..........................8. *Surnia.*

  C. Wing 12–13 in.; tail ½ wing........................1. *Asio.*

  C. Wing 5–7 in................................4. *Nyctala.*

D. Wing 14–16 in.; tail ⅔ wing........................6. *Bubo.*

D. Wing 12-13 in.; tail ½ wing........................1. *Asio.*

D. Wing 5–8 in................................5. *Megascops.*

1. *Asio.*

  * Ear tufts large, of eight to twelve feathers................

  ....*A. wilsoniànus.* (Long-eared Owl.)

  * Ear tufts small, of few feathers................

  ....*A. accipitrìnus.* (Short-eared Owl.)

2. *Syrnium nebulòsum.* (Barred Owl.)

3. *Scotiáptex cinèreum.* (Spectral Owl.)

4. *Nyctala.*

  * Bill yellow; cere not swollen...*N. téngmalmi.* (Sparrow Owl.)

  * Bill black; cere swollen............*N. acàdica.* (Saw-whet Owl.)

5. *Mégascops àsio.* (Screech Owl.)

6. *Bùbo virginiànus.* (Great-horned Owl.)

7. *Nyctea nyctea.* (Snowy Owl.)

8. *Súrnia ùlula.* (Day Owl.)

9. *Speòtyto cuniculària.* (Burrowing Owl.)

FAMILY XXVI. **STRIGIDÆ.** (BARN OWLS.)

Strix pratíncola. (American Barn Owl.)

FAMILY XXVII. **FALCONIDÆ.** (FALCONS.)

* Claws all of the same length, narrowed and rounded on the
lower side; outer toe can be used either in front or behind.
....12. *Pandion.*
* Claws of graduated lengths, the hind claw largest, the outer
front one smallest; outer toe not versatile. (**A.**)
  A. Tarsus feathered to the toes, at least in front. (**I.**)
  A. Tarsus bare for at least ⅓ of its length. (**B.**)
  B. Tarsus reticulate all around; if there are any regular scu-
tella in front, they are found only on the lower part of
the tarsus. (**G.**)
  B. Tarsus distinctly scutellate (or booted) only in front. (**E.**)
  B. Tarsus scutellate in front and behind. (**C.**)
  C. Upper tail coverts white. (**D.**)
  C. Upper tail coverts not white..............................6. *Buteo.*
  D. Wings very long (13 in. or more), more than 4 times the
length of tarsus...4. *Circus.*
  D. Wings 12 in. or less, rounded, not over 4 times the length of
the tarsus...7. *Asturina.*
  E. Toes not webbed at all at base; neck feathers sharp-pointed;
wing over 20 in. long............................10. *Haliæetus.*
  E. Toes somewhat webbed at base; wing less than 15 in. long
(**F.**)
  F. Nostril nearly circular; tail not ⅔ the length of the wing.....
....3. *Ictinia.*
  F. Nostril oval; tail over ⅔ the wing....................5. *Accipiter.*
  G. Nostril small, circular, with a conspicuous, central, bony
tubercle; upper mandible with a strong tooth and notch
back of the hooked tip.............................11. *Falco.*
  G. Nostril oval and with no inner bony tubercle. (**H.**)
  H. Tail very deeply forked.............................1. *Elanoides.*
  H. Tail not deeply forked; claws not grooved below..2. *Elanus.*

I. Tarsus densely feathered all around; wing over 22 in.........
....9. *Aquila.*

I. Tarsus with a bare strip behind; wing under 20 in............
.... 8. *Archibuteo.*

1. *Elanoìdes forficàtus.* (Swallow-tailed Kite.)
2. *Élanus leucùrus.* (White-tailed Kite.)
3. *Ictínia mississippiénsis.* (Mississippi Kite.)
4. *Círcus hudsónius.* (Marsh Harrier.)
5. *Accípiter.*
 * Wing under 9 in. long; tail square *A. vélox.* (Pigeon Hawk.)
 * Wing 9–11 in. long; tail rounded..................................
....*A. coòperi.* (Chicken Hawk.)
 * Wing 12 in. or more long............*A. atricapíllus.* (Goshawk.)
6. *Bùteo.*
 * Outer webs of primaries with white, buffy or reddish spots;
   four outer primaries notched on the inner web..................
...................................*B. lineàtus.* (Chicken Hawk.)
 * Outer webs of primaries not spotted as above. (**A.**)
 A. Four outer primaries notched on the inner web. (**C.**)
 A. Three outer primaries notched. (**B.**)
 B. Wing over 14 in. long...*B. swàinsoni.* (Swainson's Buzzard.)
 B. Wing under 12 in. long..
..........................*B. latíssimus.* (Broad-winged Buzzard.)
 C. Head and neck streaked with rusty red; tail bright chest-
   nut red.............................*B. boreàlis.* (Hen Hawk.)
 C. Head never streaked with buffy or reddish.....................
....*B. hárlani.* (Black Hawk.)
7. *Asturìna plagiàta.* (Gray Hawk.)
8. *Archibùteo.*
 * Gape less than 1½ in. wide at base.....................................
..................................*A. lagòpus.* (Rough-legged Hawk.)
 * Gape over 1½ in. wide at base...
..............*A. ferrugíneus.* (Western Rough-legged Hawk)

10. *Haliæetus* (hal-i-â-e-tus) *leucocéphalus.* (Bald Eagle.)

11. *Fálco.*

* Only one primary notched on the inner web; wing 12 in. or more long. (**B.**)

* Two primaries notched; wing under 10 in. long. (**A.**)

A. Basal joints of the toes with six-sided scales; middle toe over 1 in. long............*F. columbárius.* (Pigeon Hawk.)

A. Not such six-sided scales; middle toe under 1 in. long......
.... *F. sparvèrius.* (Sparrow Hawk.)

B. Tarsus hardly at all feathered at the upper end.................
.... *F. peregrìnus.* (Duck Hawk.)

B. Tarsus feathered in front not over half way down.............
..................................*F. mexicánus.* (Prairie Falcon.)

B. Tarsus densely feathered over half way down...................
..................................*F. rustícolus.* (Gray Gyrfalcon.)

12. *Pandion haliaëtus.* (Fish Hawk.)

FAMILY XXVIII. **CATHARTIDÆ.** (VULTURES.)

* Tail rounded; nostril large and broad.................1. *Cathárta.*
* Tail square; nostril small and narrow...............2. *Catharista.*

1. *Cathártes aùro.* (Turkey Buzzard.)
2. *Catharísta atràta.* (Black Buzzard.)

FAMILY XXIX. **COLUMBIDÆ.** (PIGEONS.)

* Tail less than ⅔ as long as the wing and rounded.................
....3. *Columbigallina.*
* Tail as long or longer than the wing and pointed. (**A.**)

A. Tarsus scutellate, entirely bare of feathers and longer than the side toes..2. *Zenaidùra.*

A. Tarsus feathered above at the joint and shorter than the side toes...1. *Ectopistes.*

1. *Ectopistes migratòrius.* (Passenger Pigeon.)
2. *Zenaidùra macroùra.* (Mourning Dove.)
3. *Colúmbigallìna passerìna.* (Ground Dove.)

FAMILY XXX. **TETRAONIDÆ**. (GROUSE.)

* Tarsus bare throughout, except possibly at the upper joint, and scutellate; nostril with a naked scale and without feathers; head not distinctly crested.................1. *Colinus.*
* Tarsus feathered about half way; tail fan-shaped; neck with a ruff of lengthened feathers...........................3. *Bonasa.*
* Tarsus feathered to the toes. (**A.**)
　A. Toes also fully feathered; no ruff or peculiar feathers on the neck...4. *Lagopus.*
　A. Toes naked, except, possibly, at the base. (**B.**)
　B. Tail more than half as long as the wing; no ruff or peculiar feathers on the neck; tail slightly rounded.....................
　...2. *Dendragapus.*
　B. Tail about half as long as the wing; no ruff; tail graduated the middle feathers much lengthened..:........6. *Pediocætes.*
　B. Tail about half as long as the wing; neck with a ruff of straight, stiff feathers............................5. *Tympanuchus.*

1. *Colinus virginiànus.* (Bob White.)
2. *Dendrágapus canadénsis.* (Canada Grouse.)
3. *Bonàsa umbéllus.* (Ruffed Grouse.)
4. *Lagòpus lagòpus.* (Willow Grouse.)
5. *Tympanùchus americànus.* (Prairie Hen.)
6. *Pediocætes* (ped-i-è-se-tes) *phasianéllus.* (Sharp-tailed Grouse.)

FAMILY XXXI. **PHASIANIDÆ**. (PHEASANTS.)

Meleàgris gallopàvo. (Wild Turkey.)

FAMILY XXXII. **HÆMATOPODIDÆ**. (OYSTER-CATCHERS.)

Hæmátopus palliàtus. (Oyster-catcher.)

FAMILY XXXIII. **APHRIZIDÆ.** (SURF BIRDS.)

Arenària intérpres. (Turnstone.)

FAMILY XXXIV. **CHARADRIIDÆ.** (PLOVERS.)

* Plumage speckled on the back; tarsus much longer than the middle toe and claw....................................1. *Charadrius.*

* Plumage not speckled on the back; tarsus but little longer than the middle toe and claw......................... 2. *Ægialitis.*

1. *Charódrius.*

* Hind toe very short......*C. squatàrola.* (Black-bellied Plover.)

* No hind toe.........................*C. domínicus.* (Golden Plover.)

2. *Ægialtis.*

* Wing 6 in. or more long; rump orange-brown....................
....*Æ. vocífera.* (Kildeer Plover.)

* Wing under 6 in. long; rump colored like the back. (**A.**)

A. Bill ¾ in. or more long.....*Æ. wilsònia.* (Wilson's Plover.)

A. Bill under ¾ in. long. (**B.**)

B. All toes distinctly webbed at base..................................
.... *Æ. semipalmàta.* (King Plover.)

B. Inner toes without webbing.....*Æ. melòda.* (Piping Plover.)

FAMILY XXXV. **SCOLOPACIDÆ.** (SNIPE.)

* Bill very long and much decurved; tarsus scutellate only in front, reticulate behind..............................16. *Numenius.*

* Bill not strongly decurved; tarsus scutellate in front and behind. (**A.**)

A. Toes only 3, hind toe wanting........................8. *Calidris.*

A. Toes 4, the hind toe present. (**B.**)

B. Eyes situated far back on the head directly above the ears; bill long, with the upper mandible thickened at the tip;

B. Eyes in the usual position; bill either long or short. **(C.)**

  C. Front toes not webbed, or, at most with one minute web. **(J.)**

  C. Front toes with at least one distinct web. **(D.)**

D. Tail more than ½ as long as wing and graduated 1 in. or more...12. *Bartramia.*

D. Tail not more than ⅓ the wing and never graduated more than ⅓ the length of the bill. **(E.)**

  E. Wing *less* than 4 in. long; bill grooved at tip...7. *Ereunetes.*

  E. Wing 4 in. or more long. **(F.)**

F. Wing about 6½ in.; bill about 1¼ in.; male with a ruff........ ....13. *Pavoncella.*

F. Bill longer in proportion to the wing. **(G.)**

  G. Wing about 4 in.; bill about 1 in.; tarsus about as long as the middle toe and claw.............................15. *Actitis.*

  G. Wing over 4½ in. long. **(H.)**

H. Bill slightly broadened at tip; bill and tarsus about equal and under 2 in. long; wing about 5 in......5. *Micropalama.*

H. Bill slightly broadened; bill over 2 in.; tarsus under 2 in.; wing 5¼–6 in.....................................4. *Macrorhamphus.*

H. Bill not broadened at tip. **(I.)**

  I. Wing 8 in. or more long; bill 3–5 in. long..........9. *Limosa.*

  I. Wing under 8 in. long; bill slender, not over 2¼ in. long; basal half of primaries not white..................10. *Totanus.*

  I. Wing usually over 8 in. long; bill stout, not over 2¼ in. long; basal half of primaries white..........11. *Symphemia.*

J. Under side of wing shows the inner web beautifully mottled... ....14. *Tryngites.*

J. Inner webs not mottled; bill longer than middle toe............ ....6. *Tringa.*

  K. Tibia naked below; crown striped lengthwise..3. *Gallinago.*

  K. Tibia entirely feathered; crown banded crosswise. **(L.)**

L. Three outer primaries narrowed and abruptly shortened...... ....1. *Philohela.*

L. Quills not narrowed; first longer than second......2. *Scolopax.*

1. *Philóhela minor.* (American Woodcock.)
2. *Scólopax rustícola.* (European Woodcock.)
3. *Gallinágo delicáta.* (Wilson's Snipe.)
4. *Macrorhámpus.*
   * Bill not over 2½ in.........................*M. gríseus.* (Dowitcher.)
   * Bill over 2½ in.; western...
     ....*M. scolopáceus.* (Long-billed Dowitcher.)
5. *Micropálama himántopus.* (Stilt Sandpiper.)
6. *Trínga.*
   * Wing 6 in. or more long; middle pair of tail feathers not
     lengthened............................*T. canútus.* (Robin Snipe.)
   * Wing less than 6 in. long; middle pair of tail feathers acute
     and abruptly lengthened. (**A.**)
   A. Bill, tarsus and middle toe of about equal length. (**D.**)
   A. Bill decidedly longer than the tarsus. (**B.**)
   B. Bill about 1¼ in.; middle toe and claw 1⅛ in.; tarsus 1 in....
     ....*T. marítima.* (Purple Sandpiper.)
   B. Bill over 1⅛ in.; tarsus over 1 in. (**C.**)
   C. Upper tail coverts mostly dusky.........*T. alpína.* (Dunlin.)
   C. Upper tail coverts white...
     ....*T. ferrugínea.* (Curlew Sandpiper.)
   D. Wing 5 in. or more; bill over 1 in.; rump and middle tail
     coverts black or dusky...........*T. maculáta.* (Jack Snipe.)
   D. Wing about 5 in.; bill not over 1 in.; upper tail coverts
     mainly white......*T. fusicóllis.* (White-rumped Sandpiper.)
   D. Wing about 4¾ in.; middle tail coverts plain dusky............
     ....*T. báirdii.* (Baird's Sandpiper.)
   D. Wing under 4 in.; tarsus ¾ in...
     ....*T. minutílla.* (Least Sandpiper.)
7. *Ereunétes.*
   * Bill about ¾ in. long..*E. pusíllus.* (Semi-palmated Sandpiper.)
   * Bill about 1 in. long......*E. occidentális.* (Western Sandpiper.)
8. *Cálidris arenária.* (Sanderling.)
9. *Limósa.*

* Tail barred with black.............*L. fédoa.* (Marbled Godwit.)
* Tail black, with white base and tip...................................
.............................*L. hæmástica.* (Black-tailed Godwit.)
10. *Tótanus.*
  * Wings 7 in. or more; legs yellow.....................................
.............................*T. melanoleùcus.* (Greater Yellow-legs.)
  * Wings 6-7 in.; legs yellow............*T. flávipes.* (Yellow-legs.)
  * Wings under 6 in.; legs dusky.....................................
..............................*T. solitàrius.* (Solitary Sandpiper.)
11. *Symphèmia semipalmàta.* (Willet.)
12. *Bartràmia longicaúda.* (Upland Sandpiper.)
13. *Pavoncélla púgnax.* (Ruff.)
14. *Tryngìtes subruficóllis.* (Buff-breasted Sandpiper.)
15. *Acùtis maculària.* (Spotted Sandpiper.)
16. *Numènius.*
  * Bill over 5 in. long; wing over 10 in.; secondaries and quills
    of wing rusty cinnamon.....................................
..............................*N. longiróstris.* (Long-billed Curlew.)
  * Bill 3-4 in.; wing 9-10 in.; crown with two broad dusky side
    stripes......................*N. hudsónicus.* (Hudsonian Curlew.)
  * Bill 2-3 in.; wing 8-9 in.; dusky crown stripes narrow.........
..............................*N. boreàlis.* (Eskimo Curlew.)

Family XXXVI. RECURVIROSTRIDÆ. (Avocets and Stilts.)

  * Hind toe present; front toes full-webbed........1. *Recurvirostra.*
  * No hind toe.................................2. *Himantopus.*
1. *Recurviróstra americàna.* (Avocet.)
2. *Himántopus mexicànus.* (Long-shanks.)

Family XXXVII. PHALAROPODIDÆ. (Phalaropes.)

  * Bill stoutish and with a flattened tip..............1. *Crymophilus.*
  * Bill very slender and not flattened..................2. *Phalaropus.*

1. *Crymóphilus fulicárius.* (Red Phalarope.)
2. *Phaláropus.*
* Membrane along the toes scalloped.......................................
...................................*P. lobátus.* (Northern Phalarope.)
* Membrane plain..............*P. tricolor.* (Wilson's Phalarope.)

FAMILY XXXVIII. GRUIDÆ (CRANES.)
Grus.
* Adult white; tarsus 11 in. or more long............................
...............................*G. americana.* (Whooping Crane.)
* Adult slaty gray; tarsus less than 11 in. long. (**A.**)
A. Tarsus over 10 in.; bill over 5 in.........................
...............................*G. mexicána.* (Sandhill Crane.)
A. Tarsus under 9 in.; bill under 5 in.............................
...........................*G. canadénsis.* (Little Brown Crane.)

FAMILY XXXIX. RALLIDÆ (RAILS.)

* Forehead with a shield-like horny extension of bill. (**B.**)
* No horny extension of bill on forehead. (**A.**)
A. Bill slender, decurved, as long or longer than the tarsus....
....1. *Rallus.*
A. Bill short, under 1 in. long, stout, not decurved..2. *Porzana.*
B. Toes with broad flap-like, lobed membranes along their edges
....5. *Fulica.*
B. Toes with little or no membraneous edges. (**C.**)
C. Nostril about ⅓ as long as the gonys................3. *Ionornis.*
C. Nostril over ½ as long as the gonys................4. *Gallinula.*
1. *Rállus.*
*. Wing less than 5 in. long....*R. virginiánus.* (Virginia Rail.)
* Wing over 5 in. long. (**A.**)
A. Plumage generally grayish....*R. crépitans.* (Clapper Rail.)
A. Plumage generally brownish or reddish.........................

2. *Porzàna.*
* Secondary quills white.........*P. novcboracénsis.* (Yellow Rail.)
* Secondaries not white. (**A.**)
 A. Wing over 4 in. long............*P. carolìna.* (Carolina Rail.)
 A. Wing under 4 in long.........*P. jamaicénsis.* (Black Rail.)
3. *Ionòrnis martínica.* (Purple Gallinule.)
4. *Gallinúla galeàta.* (Florida Gallinule.)
5. *Fùlica americàna.* (American Coot.)

<h2 style="text-align:center">FAMILY XL. ARDEIDÆ. (HERONS.)</h2>

* Outer toe shorter than the inner toe; claws lengthened..........
.... 1. *Botaurus.*
* Outer toe as long or longer than the inner; claws short. (**A.**)
 A. Bill rather slender; bill 4 times as long as its depth at
 base...2. *Ardea.*
 A. Bill stouter...3. *Nycticorax.*
1. *Botaúrus.*
* Wing 10 in. or more long..*B. lentiginòsus.* (American Bittern.)
* Wing 6 in. or less long.................*B. exìlis.* (Least Bittern.)
2. *Árdea.*
* Wing 11 in. or more long. (**C.**)
* Wing under 11 in. long. (**A.**)
 A. Bill and tarsus of about equal length; wing over 8 in.......
 ....*A. trícolor.* (Louisiana Heron.)
 A. Bill somewhat longer than tarsus; tarsus 2¼ in. or less......
 ....*A. viréscens.* (Green Heron.)
 A. Bill decidedly shorter than tarsus. (**B.**)
 B. Color pure white; plumes on back. ....
 ....*A. candidíssima.* (Snowy Egret.)
 B. Color not pure white; no plumes on back.......................
 ....*A. cærùlea.* (Little Blue Heron.)
 C. Wing 18-20 in.; color bluish......................................
 ....*A. heròdias.* (Great Blue Heron.)

C. Wing 14-18 in ; color white...
.................................*A. egrétta.* (Great White Egret.)
C. Wing 11½-14 in.; slate color; head and neck cinnamon.....
.................................*A. ruféscens.* (Reddish Egret.)
3. *Nycticorax.*
* Bill nearly as long as the tarsus; top and bottom of bill
slightly convex.........................*N. nycticorax.* (Squawk.)
* Bill much shorter than tarsus; top and bottom of bill decid-
edly convex...*N. violàceus.* (Yellow-crowned Night Heron.)

Family XLI. CICONIIDÆ. (Storks.)

Tántalus loculàtor. (Wood Ibis.)

Family XLII. IBIDIDÆ. (Ibises.)

* Head of adult wholly naked in front; plumage of upper parts
not metallic........... ....1. *Guara.*
* Head naked only in front of the eyes; plumage of upper
parts metallic or bronzy...............................2. *Plegadis.*
1. *Guára álba.* (White Ibis.)
2. *Plégadis antumnàlis.* (Glossy Ibis.)

Family XLIII. PLATALEIDÆ. (Spoonbills.)

Ajaja (i-i-u-i) *ajaja.* (Roseate Spoonbill.)

Family XLIV. ANATIDÆ. (Ducks, Geese and Swans.)

* Neck as long as the body; adult entirely white; wing 20 in.
or more long. (Swans.)...................................1. *Olor.*
* Neck shorter than the body. (**A.**)
A. Tarsus as long or longer than the middle toe without claw.
(Geese.) (**P.**)

A. Tarsus shorter than the middle toe without claw. (Ducks.) **(B.)**

B. Head crested; bill nearly cylindrical, only about as wide as high throughout; bill with saw-like teeth. (Fish Ducks.) **(O.)**

B. Head usually not crested; bill rather broadened, always wider than high near tip; lamellæ of bill bluntish. (True Ducks.) **(C.)**

C. Hind toe with a broad membraneous border or lobe. (Sea Ducks.) **(G.)**

C. Hind toe without a distinct membraneous lobe. (River Ducks.) **(D.)**

D. Bill decidedly broadened towards the tip.............6. *Spatula.*

D. Bill little, if at all, widened towards the tip. **(E.)**

E. Tail feathers broad and rounded at tip; head more or less crested...8. *Aix.*

E. Head not crested; tail feathers narrow and rather acute. **(F.)**

F. Wing over 10 in. long; bill about 2 in.; tail graduated for more than ⅓ its length... ....7. *Dafila.*

F. Wing and bill are either shorter than above, or else the tail is graduated for less than ⅓ its length................5. *Anas.*

G. Tail 4½ in. or less long, its feathers with narrow webs and stiff shafts extending beyond the webs; upper tail coverts very short. **(N.)**

G. Tail feathers with their bases well covered by the upper coverts. **(H.)**

H. Feathers of the forehead at center and feathers at the side extending on the upper mandible, so as to leave a bare portion between, ⅓ as long as the bill; no speculum.........
.... 15. *Somateria.*

H. No such extensions of both frontal and loral feathers. **(I.)**

I. Tail graduated for a less distance than the length of the bill from the nostril; nail at tip of bill less than $\frac{1}{3}$ its width at the middle.................,...........................9. *Aythya.*

I. Tail graduated for a greater distance than the length of bill from the nostril. (**J.**)

J. Bill of the ordinary duck form, neither much enlarged at the base nor with peculiar appendages at the base or sides. (**L.**)

J. Bill quite peculiar in form or appendages. (**K.**)

    K. Speculum violet; upper mandible with a side lobe at base.. ....13. *Histrionicus.*

    K. Speculum white; upper mandibles with leathery expansions on the sides; cheeks bristly.........14. *Camptolaimus.*

    K. Top of bill peculiarly bulging near the frontal feathers and usually hollowed back of the indistinct nail..16. *Oidemia.*

L. Nail of bill large but indistinct; no speculum; tail pointed; bill black and orange; nostril quite near the frontal feathers...12. *Clangula.*

L. Nail narrow and distinct; nostril near the center of the length of the bill. (**M.**)

    M. Front of nostril more than $\frac{1}{2}$ way from the loral feathers; iris yellow...10. *Glaucionetta*

    M. Front of nostril not $\frac{1}{2}$ way from the loral feathers towards the tip of bill; iris brown....................11. *Charitonetta.*

N. Nail of bill very small and the lower end abruptly bent backward; outer toe longer than the middle toe............. ....17. *Erismatura.*

N. Nail of bill rather large and not abruptly bent backward; outer toe shorter than middle toe................18. *Nomonyx.*

    O. Teeth-like serrations of upper mandible very long, sharp and strongly hooked..............................19. *Merganser.*

    O. Teeth-like serrations short, blunt and not conspicuously bent backward; crest on head high and flattened sideways ....20. *Lophodytes.*

P. Serrations on the cutting-edge of upper mandible scarcely

visible from the side at all; if visible then only at the
base; bill, feet and portions of head black.........4. *Branta.*
P. Serrations visible from the side for more than ½ the length
of bill; bill and feet pale. (**Q.**)
Q. Depth of bill at base much greater than ½ the length of the
top of bill...2. *Chen.*
Q. Depth of bill at base about ½ the length of the top of bill.
....3. *Anser.*
1. *Olor.*
　＊ Back end of nostril much nearer to the tip of bill than the
　　front corner of the eye..*O. columbiânus.* (Whistling Swan.)
　＊ Back end of nostril about midway between the tip of bill and
　　the front corner of the eye..*O. buccinâtor.* (Trumpeter Swan.)
2. *Chen.*
　＊ Plumage of adult grayish-brown..*C. cærulêscens.* (Blue Goose.)
　＊ Plumage of adult white...........*C. hyperbôrea.* (Snow Goose.)
3. *Anser álbifrons.* (White-fronted Goose.)
4. *Bránta.*
　＊ Head entirely black; side of neck with a patch of white
　　streaks..............................*B. bernícla.* (Brant Goose.)
　＊ Head partly white. (**A.**)
　A. Head mostly black; a whitish triangular patch on cheek...
　　..............................*B. canadénsis.* (Canada Goose.)
　A. Head mostly white.............*B. leucópsis.* (Barnacle Goose.)
5. *Ânas.*
　＊ Bill shorter than the head; ~~tail feathers not acute~~; belly
　　white; crown whitish. (**D.**)
　＊ Bill about as long as the head or longer. (**A.**)
　A. Speculum of wing white; wing about 11 in. long; wing
　　coverts chestnut; bill dark........*A. strépera.* (Gadwall.)
　A. Speculum of wing violet with black border; wing 10-11 in.
　　long; bill not very dark. (**E.**)
　A. Speculum of wing green; bill dark; wing not over 8 in.
　　long. (**B.**)

B. A white crescent-shaped spot on sides of body in front of wing; wing coverts leaden-gray without blue................ ....*A. carolinénsis.* (Green-winged Teal.)

B. Wing coverts and some of the shoulder feathers sky-blue. **(C.)**

  C. Male with a white crescent in front of eye; head and neck blackish lead-color.........*A. díscors.* (Blue-winged Teal.)

  C. No pure white on head; general color of male purplish-chestnut.....................*A. cyanóptera.* (Cinnamon Teal.)

D. Head and neck cinnamon-red with but little if any green.... ....*A. penélope.* (European Widgeon.)

D. Sides of head with a broad green patch........................... ....*A. americána.* (Bald-pate. American Widgeon.)

E. Male with head and neck glossy-green and a white ring below; female dusky... ....*A. bóschas.* (Mallard Duck. Tame Duck.)

E. Both sexes like the female of the preceding, but darker..... ....*A. obscùra.* (Black Duck.)

6. *Spátula clypeàta.* (Spoon-bill Duck.)

7. *Dáfila acùta.* (Pin-tail Duck.)

8. *Aix spónsa.* (Wood Duck.)

9. *Áythya.*

  * Bill decidedly wider at tip than at base. **(B.)**

  * Bill with the width at the end about equal to the width at base. **(A.)**

  A. Width of bill nearly ½ the length at top......................... ....*A. americàna.* (Red Head.)

  A. Width of bill about ⅓ the length.................................. ....*A. vallisnèria.* (Canvas-back Duck.)

B. Male with an orange-brown ring around neck; speculum bluish-gray; female chiefly brown.............................. ....*A. collàris.* (Ring-necked Duck.)

B. Speculum white in male; face white in female. **(C.)**

  C. Wing over 8¼ in. long.........*A. marìla.* (Big Scaup Duck.)

C. Wing under 8¼ in. long....*A. affinis.* (Lesser Scaup Duck.)

10. *Glaucionétta.*

* Male with the head uniformly puffy and the gloss green; spot before the eye roundish...........*G. clangúla.* (Golden Eye.)

* Male with the head somewhat crested and the gloss purplish; spot before the eye angular...
.........................*G. islándica.* (Barrow's Golden Eye.)

11. *Charitonétta albèola.* (Buffle-head.)

12. *Clángula hyemàlis.* (Old Squaw. Old Wife.)

13. *Histriónicus histriónicus.* (Harlequin Duck.)

14. *Camptolaimus* (lè-mus) *labradòrius.* (Labrador Duck.) Probably extinct.

15. *Somatèria.*

* Feathers on the side of bill extending far beyond those on the top.. ....*S. drésseri.* (American Eider Duck.)

* Feathers on the top of the bill extending far beyond those on the side...........................*S. spectábilis.* (King Eider.)

16. *Oidèmia.*

* Wing 10½ in. or more long; nostril beyond the middle of the bill; a white wing patch..
..........................*O. déglandi.* (White-winged Scoter.)

* Wing less than 10½ in long. (**A.**)

A. Nostril beyond the middle of the bill; white or whitish patches on head; no white wing patch.........................
.................................*O. perspicillàta.* (Surf Scoter.)

A. Nostril not beyond the middle; male entirely black; female with whitish on throat and sides of head....................
.................................*O. americàna.* (Black Scoter.)

17. *Erismatùra rùbida.* (Ruddy Duck.)

18. *Nómonyx domínicus.* (St. Domingo Duck.)

19. *Mergánser.*

* Nostril near the middle of bill; frontal feathers extend beyond those on side of bill...
........................*M. americànus.* (Merganser. Goosander.)

* Nostril near base of bill; frontal feathers not beyond those on side of bill..............*M. serrátor.* (Red-breasted Merganser.)

20. *Lophódytes cucullátus.* (Hooded Merganser.)

FAMILY XLV. SULIDÆ. (GANNETS.)

Súla bassána. (Common Gannet. Solan Goose.)

FAMILY XLVI. ANHINGIDÆ. (DARTERS.)

Anhinga anhínga. (Darter. Snakebird.)

FAMILY XLVII. PHALACROCORACIDÆ. (CORMORANTS.)

Phalacrócorax.

* Tail of fourteen feathers; pouch on neck notched behind.....~ ....*P. cárbo.* (Common Cormorant. Shag.)

* Tail of twelve feathers. (**A.**)

A. Wing over 12 in. long.. ....*P. dilóphus.* (Double-crested Cormorant.)

A. Wing under 11 in. long.. ....*P. mexicánus.* (Mexican Cormorant.)

FAMILY XLVIII. PELECANIDÆ. (PELICANS.)

Pelecánus erythrorhynchos. (White Pelican.)

FAMILY XLIX. FREGATIDÆ. (MAN-O'-WAR BIRDS.)

Fregáta áquila. (Man-o'-war Bird.)

FAMILY L. **PROCELLARIIDÆ.** (PETRELS.)

* Wings under 7 in. long; length under 10 in. (Stormy Petrels.) (**C.**)

* Wings over 7 in. long; length over 10 in. (**A.**)

A. Under mandible not hooked at tip. (Fulmars.)...............
....1. *Fulmarus.*

A. Under mandible hooked like the upper. (**B.**)

B. Bill over ¾ as long as the tarsus. (Petrels.)......3. *Æstrelata.*

B. Bill usually ⅔-¾ as long as the tarsus. (Shearwaters.).........
....2. *Puffinus.*

C. Claws flat, obtuse; tarsus booted, much longer than middle toe and claw; upper tail coverts white.........6. *Oceanites.*

O. Claws hooked, acute; tarsus but little longer than middle toe and claw. (**D.**)

D. Tail nearly square; color sooty-brown with white..............
....4. *Procellaria.*

D. Tail forked; color bluish or grayish with white................
....5. *Oceanodroma.*

1. *Fúlmarus glaciális.* (Fulmar.)
2. *Púffinus.*
   * Sooty-colored throughout ..*P. stricklandi.* (Sooty Shearwater.)
   * Dark above, white below. (**A.**)
   A. Wing 10 in. or less; under tail coverts mostly white ....
   ....*P. audubóni* (Audubon's Shearwater.)
   A. Wing 12 in. or more. (**B.**)
   · **B.** Brownish-ash above; upper tail coverts dark, under ones light............................*P. boreális.* (Cory's Shearwater.)
   B. Dark-brown above; upper tail coverts white, under ones dark............................ *P. májor.* (Greater Shearwater.)
3. *Æstrélata hasitáta.* (Black-capped Petrel.)
4. *Procellária pelágica.* (Storm Petrel.)
5. *Oceanódroma leucórhoa.* (Leach's Petrel.)

Family LI. **STERCORARIIDÆ**. (Jægers.)

* Tarsus shorter than middle toe and claw; wing 15 in. or more long..1. *Megalestris.*
* Tarsus not shorter than middle toe and claw; wing less than 15 in. long.................... ....2. *Stercorarius.*

1. *Megaléstris skúa.* (Skua Gull.)
2. *Stercorárius.*
   * Central tail feathers in adult projecting over 6 in. beyond the others and sharp pointed ....
   ....*S. longicaúdus.* (Long-tailed Jæger.)
   * Central tail feathers projecting less than 5 in. (**A.**)
   A. Central tail feathers broad at tip.....................................
   ....*S. pomárinus.* (Pomarine Jæger.)
   A. Central tail feathers acute at tip...................................
   ....*S. parasíticus.* (Parasitic Jæger.)

Family LII. **LARIDÆ**. (Gulls.)

* Bill more or less hooked; tail generally even; if forked, the outer feathers not narrowed or acute pointed; colors generally white, with a darker, usually grayish, mantle on back. (Gulls.) (**C.**)
* Bill not hooked, at most, slightly curved, the mandibles even; tail deeply forked. (Terns.) (**A.**)
   A. Tail with the outer feathers narrow and pointed; front toes well webbed; colors of plumage light. (**B.**)
   A. Tail with the outer feathers broad and rounded; front toes but little more than half webbed; colors dark..............
   ....7. *Hydrochelidon.*
   B. Bill dark and stout, its depth at base over ⅓ its length at top.......... ....5. *Gelochelidon.*
   B. Bill less stout, usually quite slender....................6. *Sterna.*

C. Hind toe very minute or wanting; tail slightly notched or even............... ...., ...2. *Rissa.*

C. Hind toe small. **(D.)**

D. Tail forked; bill black with a yellow tip..............4. *Xema.*

D. Tail even. **(E.)**

    E. Adults pure white; tarsus rough behind and about equal to the middle toe, without claw, in length.............1. *Gavia*

    E. Adults white with a darker mantle; tarsus not very rough and less than middle toe in length...................3. *Larus.*

1. *Gàvia álba.* (Ivory Gull.) *greater*
2. *Rissa tridáctyla.* (Kittiwake Gull.)
3. *Làrus.*

    * Wing over 14 in. long; head entirely white in adult in summer. **(C.)**

    * Wing under 14 in. long; head black or dusky in adult in summer. **(A.)**

  A. Wing over 12 in. long; tarsus longer than middle toe and claw.........................*L. atricílla.* (Black-headed Gull.)

  A. Wing under 12 in. long. **(B.)**

B. Bill red, with usually a dark band near tip.......................
.............................*L. fránklini.* (Franklin's Rosy Gull.)

B. Bill black and slender; wing about 10 in. long..................
.............................*L. philadélphia.* (Bonaparte's Gull.)

  C. Primaries pearl-gray, fading to white at tips, no black. **(G.)**

  C. Primaries with white tips and darker, dusky or black spaces near tips (in young sometimes all dark). **(D.)**

D. Dark spaces of primaries gray.......................................
...................................*L. kumlieni.* (Gray-winged Gull.)

D. Dark spaces of primaries black. **(E.)**

  E. Shafts of primaries white through the dark spaces............
.......................*L. márinus.* (Great Black-backed Gull.)

  E. Shafts dark like the spaces. **(F.)**

F. Wing over 16 in. long.........*L. argentátus.* (Herring Gull.)

F. Wing under 16 in. long..*L. delawarénsis.* (Ring-billed Gull.)

G. Wing over 16½ in. long................*L. glaúcus.* (Ice Gull.)

G. Wing under 16½ in. long ...*L. leucópterus.* (Iceland Gull.)

4. *Xéma sabínii.* (Sabine's Gull. Fork-tailed Gull.)

5. *Gélochelídon nilótica.* (Gull-billed Tern.)

6. *Stérna.*

   * Wing over 15 in. long; tail forked less than ⅓ its length.......
................................*S. techegráva.* (Caspian Tern.)

   * Wing about 15 in. long; tail forked for about ½ its length......
................................*S. máxima.* (Royal Tern.)

   * Wing less than 13 in. long. (**A**.)

   A. Head decidedly crested. *S. sandvicénsis.* (Sandwich Tern.)

   A. Head but little if at all crested; wing 12 in. or less long.
     (**B**.)

   B. Wing under 7 in. long............*S. antillárum.* (Least Tern.)

   B. Wing about 12 in. long; back sooty-black; inner webs of
     quills dusky........................*S. fuliginósa.* (Sooty Tern.)

   B. Wing 8-12 in. long; back in adult pearl-gray. (**C**.)

   C. Outer tail feathers with inner web dusky, outer web white..
................................*S. fórsteri.* (Forster's Tern.)

   C. Outer tail feathers with both webs white........................
................................*S. doúgalli.* (Roseate Tern.)

   C. Outer tail feathers with inner web white, outer web dusky.
     (**D**.)

   D. Bill red with a blackened tip; tail but little more than ⅓ the
     length of wing................*S. hirúndo.* (Common Tern.)

   D. Bill red throughout; tail over ⅔ the length of the wing......
................................*S. paradísœa.* (Arctic Tern.)

7. *Hydrochélidon nígra.* (Black Tern.)

Fᴀᴍɪʟʏ LIII. **RHYNCHOPIDÆ.** (Sᴋɪᴍᴍᴇʀs.)

Rhynchops nígra. (Black Skimmer.)

FAMILY LIV. **URINATORIDÆ**. (LOONS.)

Urinator.

* Wing over 12½ in. long; head of adult deep greenish-black.
....*U. ímber.* (Common Loon.)
* Wing about 12½ in. long; top of head bluish-ash; front of
neck blue-black...........*U. árcticus.* (Black-throated Loon.)
* Wing under 11½ in.; throat and sides of head bluish-gray; a
triangular, chestnut throat patch...............................
...................................*U. límme.* (Red-throated Loon.)

FAMILY LV. **ALCIDÆ**. . (AUKS.)

* Inner claw much longer and more curved than the others;
tarsus scutellate in front.............................1. *Fratercula.*
* Inner claw similar in size and shape to the others. (**A.**)
 A. Bill very short (½ in.) and broad; the angle of chin nearer
 to tip of bill than to the nostril; upper side of bill regu-
 larly curved.....................................5. *Alle.*
 A. Bill not so short (1 in. or more); the angle of chin nearer
 the nostril than to the tip of bill. (**B.**)
 B. Nostril overhung by a horny scale, but visible from the side;
 top of bill straight to near the tip, when it is abruptly
 curved downward...................................2. *Cepphus.*
 B. Nostril more or less completely hidden by dense velvety
 feathers. (**C.**)
 C. Tail rounded, the feathers not pointed; bill only about ⅓
 as high at base as long.................................3. *Uria.*
 C. Tail graduated, its feathers pointed; bill much deeper at
 base and much flattened sideways; wing 8 in. or more
 long; bill shorter than the head....................4. *Alca.*
1. *Fratércula árctica.* (Common Puffin.)

2. *Cépphus.*

* Greater wing coverts white throughout ....
.................................*C. mándtii.* (Mandt's Guillemot.)

* Greater wing coverts with their bases ½ black....................
...................................*C. grylle.* (Black Guillemot.)

3. *Uria.*

* Length of lower edge of bill over 1 in.............................
..................................., *U. tróile.* (Common Guillemot.)

* Length of lower edge under 1 in......................................
...............................*U. lómvia.* (Thick-billed Murre.)

4. *Álca tórda.* (Razor-billed Auk.)

5. *Álle álle.* (Dovekie. Sea-dove.)

Family LVI. **PODICIPIDÆ.** (Grebes.)

* Bill stout and somewhat hooked, its length not quite twice its
greatest depth at base.............................3. *Podilymbus.*

* Bill straight and more slender, its length more than twice its
depth at base. (**A.**)

A. Length of bird over 20 in.; neck nearly as long as the
body...1. *Æchmorphorus.*

A. Length under 20 in.; neck much shorter than the body....
.... 2. *Colymbus.*

1. *Æchmórphorus occidentális.* (Western Grebe.)

2. *Colymbus.*

* Bill about as long as the head; wing over 7 in. long.............
...................................*C. holboëlii.* (Red-necked Grebe.)

* Bill much shorter than the head; wing under 6 in. long. (**A.**)

A. Bill flattened sideways and thus higher than wide at base...
...................................*C. aurítus,* (Horned Grebe.)

A. Bill wider than high at base..*C. nigrícollis.* (Eared Grebe.)

3. *Podilymbus pódiceps.* (Pied-billed Grebe. Dab-chick.)

GLOSSARY.

Acute. Sharp-pointed, as the tip of a feather.

Barred. With cross-bands of color.

Bars. The bands of color extending across quills or feathers.

Booted tarsus. One having its front covered by a continuous scale nearly to the toes.

Bristles at rictus. Bristle-like hairs found at the corners of the mouth and extending downward.

Cere. A special skin-like covering over the base of the upper mandible, extending beyond the nostrils.

Commissure. The line which marks the closing of the mandibles; the line of the closed mouth.

Compressed. Flattened sideways.

Convex. With outward-bulging outline.

Coverts. Small feathers which hide the bases of the quills, as the upper and lower coverts of the wings and tail; also sometimes used for the covering feathers of the ears.

Crest. The lengthened feathers on the tops of the heads of some birds.

Crissum. The under tail coverts.

Crown. The top of the head, not including the portion next the bill, called the forehead,

Culmen. The ridge or central line of the upper mandible.

Decurved. Curved downward.

Depressed. Flattened above and below.

Ear tufts. Peculiar tufts of feathers found on some owls in the region of the ears.

Elevated toe. A hind toe which has its base attached to the tarsus above the level of the front toes.

Exserted. Exposed or extending beyond.

Flexible. Bending readily; not stiff.

Forehead. The part, usually feathered, just above the bill next the crown.

Forked tail. One deeply notched.

Gape. The opening of the mouth.

Gonys. The central ridge of the lower mandible from the point to where two branches form the rhami.

Graduated tail. One in which the middle pair of feathers are longest and each successive pair outward are gradually shorter.

Lamellæ. Plate-like processes found inside a duck's bill.

Lateral. At the side, as the outer tail feathers or the side toes.

Lobate toes. Furnished with membraneous flaps along the sides.

Lobes. Rounded projections.

Lore. The space between the eye and the bill.

Mandibles. The upper and lower jaws of birds.

Nail of bill. The peculiar added horny part at the tip of the upper mandible of ducks.

Nasal. Pertaining to the nostril. *Nasal groove.* The groove in which the nostril is found.

Nostrils. The openings, usually near the base, of the upper mandible.

Notched bill. One having a nick, usually near the tip of the upper mandible.

Obtuse. Somewhat angular but not sharply so; forming less than a right angle.

Ochraceous. An orange-brown or dull, deep buff.

Olivaceous. A greenish-brown color like that of olives.

Pectinated. Having comb-like notched projections, as the middle claw of herons.

Plumage. The feathering in general.

Plumes. Peculiar ornamental feathers found on the heads and backs of some birds.

Primaries. The outer nine or ten quills of the wings; those fastened to the outer joint or "hand wing." *First primary.* The first and under one of the primaries. In a mounted bird the feathers of the body should be carefully pressed back while the bird is inverted in order to determine its length.

Quills. As generally used in bird books, the stiff pen feathers of the wings, sometimes restricted to the primaries and sometimes made to include those of the tail also.

Reticulate tarsus. One covered with small irregularly-shaped scales, and thus marked with a network of lines.

Rictal bristles. See bristles at rictus.

Rictus. The back portion of the gape of the mouth, often and more properly made to include the whole gape.

Rump. The portion of the back just in front of the upper tail coverts.

Scutellate tarsus. One covered with square scales, often in front only, sometimes in front and behind.

Secondaries. The long pen feathers of the second joint or forearm of the wing; those next the primaries.

Serrations. Notches or saw-like edges, as those on a duck's bill.

Speculum. A brightly-colored spot on the wings of ducks, etc., usually consisting of the secondaries.

Tarsus. The joint of a leg just above the toes; the joint extending from the toes to the true heel.

Tertials or *tertiaries.* Properly the inner quills of a wing, those growing to the elbow or the humerus; often applied to any inner secondaries which are peculiar in length or color.

Tibia. The joint of the leg above the tarsus; the "drum stick" joint.

Truncate. With a square tip.

Versatile toe. One that can be used either in front or behind, as the outer toe of owls.

Web. The thin membrane between the toes of ducks and many snipe; also applied to the lateral halves of feathers.

Wing bars. The stripes formed by the peculiarly-colored tips of the wing coverts.

INDEX.